# ALL ABOUT SCARECROWS

by
Bobi Martin

Photographs by
Bob Allen and Teresa Willis

Cover design and illustrations by Diane Wilde

Tomato Enterprises
Fairfield • Sacramento

First Edition
©1990 Bobi Martin


Published by:
Tomato Enterprises
P.O. Box 2805
Fairfield, CA 94533
(707) 426-3970

ISBN 0-9617357-5-9

Library of Congress Catalog Number: 90-70958


All rights reserved.
Printed in the United States of America.

This book is dedicated to anyone
who has ever harbored a secret desire to build a scarecrow.

## ABOUT THE PHOTOGRAPHS

The author and publisher wish to express special appreciation to the Nut Tree, of Vacaville, California, for the use of the photographs in this book. The pictures were taken over several different seasons of the Nut Tree's annual Great Scarecrow Contest. Many were shot by Bob Allen for the Nut Tree's archives. Others were taken by Teresa Willis specifically for this book. Our thanks to both photographers, and to the staff and management of the Nut Tree.

# CONTENTS

### CHAPTER 1

All About Scarecrows:                     Page 9
History, folklore and modern day scarecrows

### CHAPTER 2

Simple Scarecrows:                       Page 20
Sun reflectors, windchimes, whirlygigs

### CHAPTER 3

Traditional Scarecrows:                  Page 30
Building from the frame up

### CHAPTER 4

Scarecrow Contests:                     Page 54
Where to see other scarecrows
or show off your own

PHOTO COURTESY OF THE NUT TREE

"Dorothy leaned her chin upon her hand and gazed thoughtfully at the Scarecrow. Its head was a small sack stuffed with straw, with eyes, nose and mouth painted on it to represent a face... She was surprised to see one of the eyes slowly wink at her."
from The Wizard of Oz, by L. Frank Baum

CHAPTER 1

# ALL ABOUT SCARECROWS

he straw man from the book and movie version of *The Wizard of Oz* is probably America's most famous scarecrow. But, while many of us think of them as distinctly American, scarecrows actually trace their roots to cultures around the world.

As long as people have raised crops for food, they have needed to keep birds and other pests from devouring the plants before harvest. Birds can gobble up newly sown seed, and tear out tender green sprouts barely poking through the soil. Throughout history, people have sought ways to protect their crops from these pesky thieves.

Thousands of years ago, Egyptian farmers along the Nile River stretched fine mesh nets over wooden frames, and set them in their wheat fields. Then, the farmers wrapped themselves in long white scarves and hid. When the birds showed up, the farmers bolted toward them, shouting and flapping their long scarves. As the frightened birds flew into the nets, the farmers caught them and stuffed them into sacks. While this process seems totally unrelated to the straw-stuffed fellow who befriended Dorothy, it is history's first recorded scarecrow.

Years later, the Greeks carved fig wood statues of their harvest god, and placed them in the fields to watch over their crops. The statues would hold a wooden club in one hand to look fierce, and a sickle in the other for good luck at harvest.

The Romans also put up statues of their harvest god to protect crops. A farmer placed offerings of flowers or fruit in the statue's apron whenever he gathered produce. During their occupation of the lands that would become England, France and Germany, the Romans taught those farmers to make these statues, also. Over time, these countries developed their own unique scarecrows.

In Germany, farmers built wooden witches, and placed them in the fields at the end of winter. They believed the witches would draw the evil spirit of winter into their bodies so that spring could come again. When planting time arrived, the witches were left in the fields to keep birds away from new seedlings. Eventually, these wooden figures came to be called *Vogelscheuchen,* or bird shooers. Potato sacks were often used for the head, and pieces of metal dangled on strings from the end of the scarecrow's arms. The clanking of the metal moving in the wind scared the birds.

An old German story tells about scarecrows who became warriors. A small kingdom discovered it would be attacked by a larger, more powerful one at daybreak. To protect itself, the kingdom needed many more warriors. But where could they be found on such short notice? Someone suggested the scarecrows. Far into the night, the desperate people gathered every scarecrow they could find. They dressed the stick figures for battle and placed them in the fields between live warriors. Just before daybreak, the attacking army approached. In the dim light, and from a distance, the scarecrow warriors looked real. Thinking they were greatly outnumbered, the attackers turned and fled!

On the other side of the world, Japanese farmers have also tried many types of scarecrows. Since very little of Japan's land can be used for farming, each plant is especially important. An early method of scaring birds was to tie bundles of oily rags and fish bones to tall sticks placed about in the rice paddies. When the farmers set the bundles on fire, the stench kept the birds and animals away. It kept people away, too! This scarecrow was called *kakashi* which means bad smell.

Over time, all Japanese scarecrows were called kakashi. Some of these included cloth streamers fluttering in the wind and bamboo sticks hung as windchimes. Japanese farmers also tied shiny bits of metal or glass to ropes strung from poles around the rice fields. The shiny

*According to legend, Sohodo-no-kami, the Japanese harvest god, would inhabit a human-looking scarecrow each season.*

objects would reflect the sunlight, which frightened off birds. The Japanese also made scarecrows that looked like people, sometimes draped in peasant clothing.

The ancient Japanese revered a scarecrow god they called Sohodo-no-kami. Each spring, they believed, he would leave his home high in the mountains to protect the new rice plants. According to legend, Sohodo-no-kami would choose a human-looking scarecrow to inhabit until fall. Farmers often left offerings of fresh fish and rice cakes at the feet of their scarecrows in hopes the god would bless their fields with his presence.

At harvest time, Sohodo-no-kami was thanked for his protection with a ceremony called "the ascent of the scarecrow." The farmers gathered every kakashi from the fields, and put them all in a pile. Special rice cakes were offered for the god to eat on his journey home, then the scarecrows were set afire.

In 17th century England, farmers tied feathers to long pieces of string which they stretched across their fields. The motion of the feathers dancing lightly in the wind would help shoo away the birds. At planting time, a farmer often had a young boy follow him as he worked. The boy would jump about and make noise, sometimes firing off a gun, which generally kept birds away.

Before European colonists came to North America, Native American tribes had many forms of scarecrows. The Seneca Indians used birds themselves. At planting time, the Seneca soaked some seeds in a poisonous mixture made from herbs. They spread the poisoned seeds around the field when the good seeds were planted. Birds that ate the poisoned seeds flew dizzily about, frightening other birds away.

Other Indian tribes would post human guards in the fields, to scare off birds as necessary. The Creek Indians selected families who moved into special huts built between the corn fields. After harvest, the families returned to their village.

Other Native American scarecrows included strips of cloth, animal skins and bones hung from rawhide thongs. These would move in the breeze and scare off birds. Farther south, in the highlands of Mexico, the Indians would place carved wooden hawks on top of pillars to guard their farmlands.

*This cornfield song-and-dance man
comes with his own chorus line of crows.*

Early British colonists in the United States stood guard over their crops, as they had often done in England. But by the 1700s, they were cultivating larger fields, and couldn't spend much time watching for birds. So, they built scarecrows dressed in cast-off clothing, with heads made from large turnips or other gourds.

The Pennsylvania Dutch, who came from Germany, brought with them the tradition of people-type scarecrows. They often built them in twos: a *bootzamon* to protect the corn or cherry trees, and his wife, a *bootzafrau,* who guarded the chicken coop to protect the baby chicks from crows and hawks. (It was rumored that the bootzamon could move about at night. Over time, "bootzamon" became "bogeyman." People venturing out after dark were sometimes warned to "Watch out for the bogeyman!")

Throughout the world, as farms increased in size and technology developed new methods, farmers turned to chemicals to control birds and other pests. Traditional scarecrows all but faded out. But in recent years, scientists have discovered that these substances stay on food longer than was previously thought. There is now concern that chemicals which kill bugs and birds might harm humans as well.

This fear of pesticides has prompted renewed interest in fending off birds without the use of chemicals. A variety of contraptions that set off loud explosions have been developed in the United States and Great Britain. One problem is that sounds that disturb birds can also disturb people. Another problem is that birds can memorize patterns of explosions, and raid the fields in between the bursts of sound! Scientists have also invented a number of mechanical devices that move in ways that frighten birds. But farmers must change the position of the machines every few days. Otherwise, the birds will get so used to the movements they'll totally ignore them.

Commercial farmers in the United States will probably never return to the days of using people-type scarecrows like Dorothy's friend in *The Wizard of Oz*. But making these traditional figures is becoming popular once more as an art form, and as a colorful and nostalgic addition to backyard gardens and orchards. Although scarecrows actually meant to scare birds would really be needed in the spring, there has been a shift to making scarecrow-type figures in the fall. These harvest time works of art often incorporate pumpkins, apples, dried cornstalks and other autumn materials. Nobody's really trying to scare crows with these figures—just to celebrate the Halloween season and have a little fun! Many times they are displayed along with bales of hay and jaunty jack o'lanterns.

One of the best places to see a wide variety of scarecrows is at the Great Scarecrow Contest held each October at the Nut Tree, in Vacaville, California. This contest offers a $1,000 prize to the best example of a traditional scarecrow, and draws entries from near and far. (You can read more about this contest in Chapter Four.)

There isn't any wrong way to make a scarecrow. You can make them out of almost anything, and put them just about anyplace you like. Let's build one and see!

*Burlap, buttons and weeds
are the main ingredients of this traditional scarecrow.*

*A Nut Tree contestant puts a finishing touch on her traditional scarecrow.*

*This wispy-looking young lady was created from a pillowcase, wig, scarf and dress. Simple and lovely.*

*Natural materials
and Father Time's famous scythe.*

*A rocking horse and plastic pumpkin face give a child-like joy to this smiling scarecrow.*

CHAPTER 2

# SIMPLE SCARECROWS

et's scare some crows! Several quick and easy ways let you harness the sun and wind. The Japanese, Navajo and many American pioneer farmers used these simple methods to keep birds away. These devices don't look anything like their traditional people-type cousins, but they get the job done just the same.

## SUN REFLECTORS

Sun reflectors are a good way to protect individual fruit trees or small garden plots. Their dancing flashes of light will act as a "keep out" sign for birds.

What you need:  scissors
pliers
sticks or small dowels
string, yarn or fishing line
reflective things (like bits of glass, shiny metal, tin foil, small mirrors, metal belt buckles, shiny jewelry.)

First decide which shiny things you'll use. If you use glass, pick pieces with dull edges or wear sturdy gloves to protect your hands. With a piece of string or yarn, tie each piece of glass as if you were gift-wrapping a package, so it can be hung.

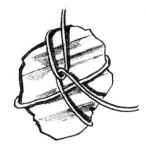

*Tie pieces of glass like a gift package.*

Disposable aluminum pie tins can be cut into many unusual and interesting shapes. If the sides feel sharp, use pliers to turn the edges under and crimp them. To hang your shapes, carefully poke a small hole about half an inch from the top of each one. Put a length of string through the hole and knot it firmly. Choosing a variety of different sizes and shapes will make your sun reflector more interesting.

Besides glass or metal, you can also cut shapes from cardboard and wrap them with aluminum foil. Remember to poke a hole near the top of each shape so it can be hung.

There are two easy ways to build your sun reflector: like a mobile or like a wind chime.

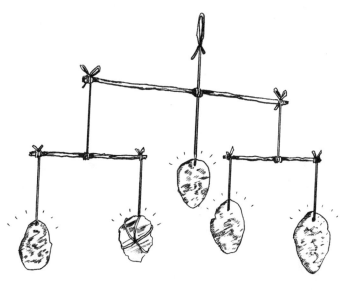

*Sun reflector mobile*

Mobiles: For this design, start with your longest stick, since everything else will hang from this one. Then choose two smaller sticks, each about a third the length of the support stick. Tie string or fishing line around the middle of each of the shorter sticks, and attach them to the support stick. For good balance, they should be tied an equal length from each end. (See illustration previous page.)

Now tie your reflecting pieces to the short sticks. You can either tie one shiny piece from the middle of each short stick, or hang two by tying one from each end. An additional reflector can hang from the middle of the support stick.

To hang the mobile, tie a loop at one end of a length of string and tie the other end to the middle of the support stick. Slip the loop onto a tree branch.

Windchime

Wind chimes: A faster way to make a sun reflector is to simply tie all of your shiny pieces to one long stick. You can hang each shape on the same length of string, or vary the lengths to create the effect that most pleases you.

To hang a wind chime, tie a length of string to one end of the support stick, then tie the end of the string to the other end of the wind chime. Now hang it from a tree and watch the birds disappear!

## STREAMERS

Brightly colored streamers fluttering in the wind are a cheerful addition to your garden. Their gentle motion also helps keep birds away. They are easy to make from materials you probably have around the house.

> What you need:  stapler or tacks or hammer and nails
> scissors
> crepe paper, ribbon or cloth
> a tall pole (or a fence or a tree)

For best results your streamers should be about two inches wide and about three feet long. You can experiment with wider or narrower pieces and with longer or shorter lengths to find which works best for you.

Crepe paper doesn't cost much, but cloth or ribbon will last longer. You may even have fabric scraps at home that you can use. Choose a lightweight cloth so that even a gentle breeze will move it.

If your garden is fenced with wire, tie one end of each streamer to the top wire. You can anchor streamers to a wooden fence with tacks, nails or staples. If your garden isn't fenced at all, drive a wooden pole into the ground and attach the streamers to it. The size of your garden will determine how many stakes you want.

How tall you want the stakes depends upon your crops. Corn or sunflowers need taller stakes than pumpkins or tomatoes. You want the streamers to flutter a foot or two above the crop.

*Streamers are a colorful addition to your garden.*

There are two ways to anchor a stake or pole. One way is to trim the bottom of the pole into a point, and then hammer the pole into the ground. An easier way is to dig a narrow hole, 12 to 18 inches deep, put the stake into the hole, and pack it tightly with dirt and rocks.

## NOISEMAKERS:

You can add sound effects to your garden by making a noisemaker.

What you need:  a stick or dowel
string, yarn or fishing line
"noisy" items such as wooden beads, broken pottery, tin cans, or bamboo sticks.

Noisemakers are made just the way the "wind chime" sun reflector is made. However, instead of mirrors, you want things that will bang around and make noise when they blow in the breeze! Tie the items to the string, and tie the strings to your support stick. Hang it from a tree, or from a nail on the side of a pole. (See illustration page 22.)

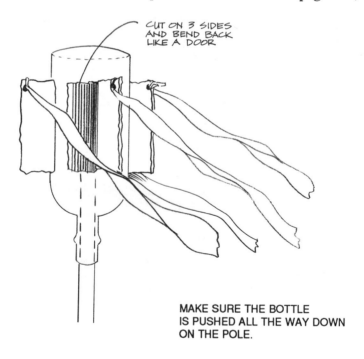

CUT ON 3 SIDES AND BEND BACK LIKE A DOOR

MAKE SURE THE BOTTLE IS PUSHED ALL THE WAY DOWN ON THE POLE.

**Whirlygig**

## WHIRLIGIGS:

Powered by the wind, these are easy to make and allow many creative touches. The movement and noise are what keep birds away.

What you need:
- a pole or stake
- a pen or felt tip marker
- scissors
- an empty plastic bleach bottle (washed!)
- OR a liter-sized plastic soda pop bottle

Strip any paper labels from your plastic bottle and wash it out well. (Unwashed soda bottles can attract ants.) With the pen or felt tip marker draw several tall rectangles around the circumference of the bottle. Try to leave the same amount of space between each rectangle. They should look like doors. Actually, they will form "wings" that catch the wind and make the bottle spin.

Use one end of the scissors to punch a hole near the top of each rectangle, then cut the top, bottom and LEFT SIDE ONLY of each wing. Bend the wing toward the right, like an open door.

To mount the bottle, put it spout-side down on the top of a thin pole. As the whirlygig rests lightly on the tip of the pole, the wind will catch the wings and spin the bottle around. For extra fun, paint designs on the bottle or attach short streamers to the wings.

## PIE PLATES AND TIN CANS

Other "recycled" materials that make good scarecrows are aluminum pie plates and tin cans. You could make an entire scarecrow by tying together several pie tins, using smaller ones for the hands. If you like, spray paint some clothes on. This fellow could hang from a tree, doing double duty as a sun reflector and a windchime!

A variation would be a figure made from an assortment of empty soup cans. Remove the bottom of the cans and connect them "tube style" on pieces of rope. You can leave the labels on the cans for an interesting effect, or take them off. Tie the arms and legs to the body. For the hands and feet, use aluminum pie tins, or tie a can sideways, putting a string through the can and tying it back to itself.

Metal lids and cake pans "bake up" a scarecrow
that moves with the wind and reflects flashes of sunlight.
Two canning jar rings make perfect ears.

*Tin cans of assorted sizes, tomato frames and other metal pieces make a strong visual statement. All that shiny metal also makes this a good sun reflector.*

*This metal sculpture evokes so-called "primitive" art of other cultures.*

*This one needs no stuffing—a wonderful use of "found" material.*

CHAPTER 3

# TRADITIONAL SCARECROWS

Traditional people-type scarecrows are easy and fun to build. The kinds of materials you can use are almost limitless. All sorts of left-overs from other projects, such as styrofoam train tunnels, funnels, springs, and slinkies can be used to build your scarecrow.

Your scarecrow can be fierce, goofy, spooky or friendly. It can be a reminder of the past, something out of the future, or anything you can dream up.

| | |
|---|---|
| Tools you will need: | hammer and nails<br>scissors<br>wire cutters<br>stapler |
| For the frame: | corn stalks, PVC pipe, chicken wire or pieces of wood (2 x 4's or 1 x 2's) |
| Clothes: | just about anything will do; also hats, gloves and other accessories |
| Face and hair: | felt tip markers or poster paints<br>pillow case, burlap sack or fabric<br>large aluminum pie tin, or mask<br>rag mop, wig, yarn, fabric scraps |

*This traditional American scarecrow has a pleasant homespun aura about him. Safety pins attach the gloves to his sleeves, and a rope secures his waistline.*

Stuffing: newspaper, straw, rags, leaves and vines

Other items: string, twine or thin wire
safety pins
pieces of styrofoam

Since scarecrows were traditionally built from castoffs the farmers had on hand, don't be afraid to substitute items on the list above. Be inventive — find new uses for old leftovers!

What materials you use and how you dress your scarecrow will determine the personality your creation takes on. Some people start with a definite idea in mind and look for items to carry out that theme. Others prefer to acquire various materials, and then let the scarecrow develop its own personality as they go along. Your scarecrow might reflect hobbies or interests you have. If you are a baseball fan, how about a pair of scarecrows playing catch? Dress them in baseball caps, and a castoff team shirt, if you have one. How about gloves and a wiffle ball?

*A Weber grill forms the body of this whimsical airplane, and supports its jolly pilot as well.*

*Is there any doubt
that the builder of this scarecrow is an avid skier?*

*Two generations of scarecrows guard this corn field.
Fancy shoelaces tie the sleeves closed,
and bright garden gloves fashion the hands.*

*An old sock for one hand, a mitten for the other, bright patches and other "found" objects create lots of visual interest. Streamers or reflective pieces could dangle from the wire jump rope.*

Let your imagination run freely. Perhaps a chunk of styrofoam from a packing box looks like a robot face to you. Hmmm, add a pair of old drawer knobs for eyes and a couple of tangled or stretched out slinky toys for arms, and you're on your way.

THE FRAME: The basic part of any scarecrow is the frame. What you want from your scarecrow determines how simple or elaborate the frame will be. Most scarecrows are built to stand in the garden and do nothing more than keep birds away. For these, the simple "T" frame is best.

On the other hand, if the scarecrow is to have positionable arms or legs, the frame must be built to allow moveable parts. This takes more time, but allows endless possibilities.

Let's start with the basic "T." You need two pieces of wood. These can be 2 x 4's or 1 x 2's (which are a little smaller and less sturdy). It doesn't matter if the wood has knots in it, has paint on it, or has been used for something else before. This is the scarecrow's skeleton and won't be seen.

The longer piece of wood should be about five or six feet long. This is the supporting piece, which stands upright and holds everything else in place. The shorter crosspiece becomes your scarecrow's shoulders. When choosing your crosspiece, it helps to have an idea of what your finished product will be. For example, if you want a Paul Bunyon-type scarecrow, choose a longer crosspiece, perhaps two and a half feet. A more delicate "lady" scarecrow would need smaller shoulders, barely two feet wide.

Measure down eight to ten inches from one end of the longer board. Center the shoulder board at this spot and nail it firmly in place. The part of the support board that sticks up above the crosspiece is where you will attach your scarecrow's head.

Remember that wood is not the only material that can be used for a frame. Dried cornstalks or sunflower stalks will work fine, though they won't last as long. You can bind three or four of the stalks together with string at the top, bottom and middle. For the crosspiece, tie two or three stalks. Don't bother to strip off the leaves, since they will add bulk and help stuff your creation. Attach the crosspiece to the

*The delicate use of natural materials
makes for a truly beautiful Harvest Queen.*

main piece by lashing the two together with string. To make arms, bend the stalks at one or both ends, or break them off at the length you prefer.

You can create a scarecrow with moveable arms and legs by using short pieces of wood and nailing them together at the joints. Lay the two pieces of wood end to end. Use long nails with big heads. Hammer three nails part way into each piece of wood. Next, bend the nails and hammer the heads into the opposite board. (See illustration below.) For added security, hammer two nails across each set.

Another way to pose a scarecrow is to use several shorter pieces of wood for the frame, instead of two long pieces. Overlap the ends of the boards and nail them together in the exact position you want the scarecrow to hold. This makes your scarecrow look like he can move, and you can take advantage of shorter pieces of wood.

Metal poles, or plastic tubing like PVC pipe may also be used for frames. You could also try chicken wire, which can be bought by the roll at building supply and feed stores. It is easily cut with wire cutters and can be shaped into interesting and unusual frames. Wear sturdy gloves since the cut ends of the wire are sharp. Shape the wire as you wish, then attach the edges by twisting the cut ends together, or tie the ends with twine or wire. Attach the head, arms and legs the same way. You may still want to use a main support stick to hold up your scarecrow and to keep it in place in your garden. Staple the chicken wire to the board, or tie it with twine or wire.

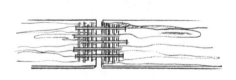 

*Connect pieces of wood to create a pose.*

*A fanciful, burlap ballerina*

CLOTHES: Let your mind take flights of fancy when it comes to dressing your scarecrow. You can add patches, funny hats, jewelry, even fake birds for that one-of-a-kind look. The clothes you choose can be garage sale bargains or closet castoffs.

If your scarecrow is to be strictly traditional, old jeans or coveralls are the usual choice for pants. Add a flannel work shirt, a pair of garden gloves and a pair of old boots and there you are. A straw hat and old neckerchief complete the outfit.

Of course, nothing says your scarecrow has to wear work clothes. Maybe there's an out-of-date suit in the closet. Why not make a "Dapper Dan"? Tuck a stuffed bird into the front pocket instead of handkerchief.

Perhaps you'd prefer a lady scarecrow. What will she wear? Jeans and a sweatshirt or a dress and hat? What if the hat was really a fake bird's nest, complete with occupants? Try a long skirt and bonnet for a prairie-girl look.

Lots of rain in your area? Give your scarecrow an umbrella, raincoat and boots. Too little rain? Turn the umbrella into a parasol using ruffles and lace, then dress the scarecrow in shorts or a sun dress.

STUFFING: Most scarecrows require some stuffing. You have lots of materials to choose from. Newspaper works well and is usually free. Just crumple it up and stuff it in. Rags, scrap fabric, and cotton batting are other good choices. Straw is cheap and can be found at feed stores. (You can use the leftover straw for mulching your garden.) Dry leaves and vines will also work.

Lay your frame on the ground. Starting with the scarecrow's pants or overalls, slip one leg over the bottom of the support stick. Let the other leg hang free. Stuff both legs. You can tie a string around the bottom of the pant legs to keep the stuffing from falling out. (Obviously, you can't stuff a skirt or a dress. Let these flutter in the breeze.)

A dress or overalls will stay on the frame, but skirts and pants need help. Run a belt or piece of rope through the belt loops and cinch it tight. Or make suspenders from ribbon or rope. (Tie one end to a back belt loop, then cross over the shoulders and tie the other end to the opposite front belt loop. Repeat for the other suspender.)

*Scarecrows can parody personalities in the news.*

Slip the sleeves of the shirt over the crosspiece. Button the shirt halfway and tuck it into the pants. Stuff the shirt and sleeves, then button all but the top two buttons. You need to leave these open to attach the head.

You can sew gloves to the ends of the sleeves, or use safety pins. You may stuff them or not as you wish. Pieces of wire coat hanger can be put into the fingers to bend them into position, such as to hold a rake. Add shoes by tucking the pant legs into them. The shoes don't need to be stuffed.

FACES AND HAIR: Making the head can be one of the most fun and creative parts of putting your scarecrow together. When choosing materials for the head, keep in mind what your finished scarecrow is going to look like. You wouldn't want to put a scary Halloween mask on a fancy lady scarecrow. And chances are you won't want to spend time stitching a great crewel yarn face for a scarecrow that's meant to be scary or silly.

An aluminum pie tin is a fast, easy way to make a face. Bring the pie tin several inches down on the support stick. Staple once or twice at the top and again at the bottom, or a strong breeze might cause the head to bend and break off.

*A sunflower gone to seed forms the basis for this smiling face. Glue holds the felt features in place.*

This is a good example of
using leftover pieces of wood.

*"Crowmen Miranda" shows that a sturdy frame need not be bulky. The jaunty pose and skillful use of natural materials make for a stunning creation.*

*Modeling wax can offer facial features with a flair. Note the peanut shell necklace, and elaborate headress made of gourds and Indian corn.*

*Old burlap feed bags are an ideal material for heads.*

Markers, paints or even stickers work well for facial features on pie tins. Hair for this type of head will have to be lightweight. Try strips of cloth or yarn. Starting at the top center, take two or three pieces of yarn or cloth and fold them over the edge of the pie tin about an inch. Then staple or glue them in place. Continue until you have as much hair as you want. The length is up to you.

You can recycle items like old plastic jugs or pieces of styrofoam for unique and interesting heads. Use dried flowers, vines, or bits of straw for the hair. You can make a soft-sculptured head by stuffing batting into a nylon stocking. Or try papier-mâché. (Spray clear lacquer over the head when it's dry to "weather-proof" it.)

Pillow cases, burlap sacks and other lengths of fabric can also make good heads. It is best to put these onto the frame first, and stuff them, before making facial features. The exception would be if you plan to stitch the eyes, nose or mouth with yarn or embroidery floss.

The head may be stuffed with the same material as the body. Stuff it partway, and then place it over the neck piece (the top part of the support stick). You may need someone else to help hold the head while you finish stuffing it.

*This "Raggedy Ann" scarecrow has a rope and yarn hairdo, burlap face and features made of felt.*

Wrap string or twine around the open end of the head, but leave two or three inches of fabric below the twine. Tie once at the front of the neck and again at the back. Let the ends of the twine hang down inside the shirt. Tug on the ends of the fabric to make sure the head is firmly in place.

It isn't necessary to hide the string. If you don't want it to show, just put a scarf or neckerchief over it. Or, you might be able to button the shirt up enough to cover the string.

To give the face some shape, just scrunch the stuffing up a bit where the nose should be. Now put on the eyes, mouth and other features with markers or paint. A wig or rag mop will make good hair, or you could draw or paint on the hair if you'd rather. Hats can be tied, glued or pinned to the head.

PLANTING YOUR SCARECROW: An easy way to "plant" a scarecrow in your garden is to dig a narrow hole about a foot deep. Put the end of the support stick into the hole, then fill it up with dirt. It's best to toss in a few shovelfuls of dirt at a time and stomp firmly, repeating the process until the hole is completely filled. It will be much easier if someone else holds the scarecrow upright while you fill the hole.

Another way to secure your scarecrow is to hammer the stake into the ground using a sledge hammer. Cutting the bottom of the stake into a point will help it go into the ground more easily.

Once the scarecrow is standing on its own you can add final decorative touches. Perhaps you'll want to attach a butterfly net to one hand, or have the scarecrow hold a rake. How about a brightly colored pinwheel in each hand?

Now step back and admire your creation!

*The wire frame lends a ghostly appearance to this cornfield bride.*

*Two young visitors admire this pumpkin-headed ballerina.*

*This interpretation of the scarecrow from "The Wizard of Oz" achieves an eerily human-like face.*

This creative sculpture uses tools and metal scrap
to evoke a pensive mood.

Recycled magnetic tape makes perfect feathers for this "scarecrow crow," which stands guard over a human effigy.

CHAPTER 4

# SCARECROW CONTESTS

Scarecrow contests date back to at least the 1800s, when Zuni children of the American southwest competed with each other to see who could make the most unusual scarecrow.

Early farm communities in the mid-western United States sometimes held scarecrow contests as well. Parents and children built scarecrows together to represent their farm and family. Awards were given for the most colorful and most unusual ones. These competitions were a way for people to express creativity, and join their neighbors in a little fun.

Scarecrow contests are still fun today, and the Guinness Book of World Records has given rise to scarecrow building in a big way. In the 1987 edition, the honor for building the world's largest scarecrow went to a youth club in Creil, Netherlands. The group's creation, called "Peuntje," stood over 41 feet tall, and was 35 feet wide. By the 1990 edition of the Guinness Book, that record had been surpassed by a team of 20 men from North Yorkshire, England. Their scarecrow towered at a whopping 90 feet, 1 inch tall, and had an arm spread of 65 feet!

The Guinness Book doesn't list the largest scarecrow competition, but that honor would go to the Nut Tree's annual Great Scarecrow Contest. The Nut Tree, a restaurant and shopping complex in Vacaville, California, features the contest as part of a month-long harvest fair celebrated each October. Over 250 scarecrows are on display, and

*The frame under this award-winning owl is chicken wire.
Small bundles of straw are tied with twine and put together
so closely that the frame is invisible. Burlap covers the main body,
adding texture and interest.*

live scarecrow mimes roam the grounds. There are also food booths, games and pumpkin-carving demonstrations. The event draws up to 20,000 people a day for the festivities.

All of the scarecrows entered in the contest are displayed in a special "gallery" in a corn patch. Corn stalks, bales of hay and tons of pumpkins provide a picturesque backdrop. The contest is divided into youth and adult categories, and then divided again for scarecrows built by one person, or by a group. Hundreds of dollars in cash prizes are awarded in each of the various divisions. People enter the contest for fun, and also with an eye toward winning a prize for their efforts!

A goal of the contest, according to Nut Tree officials, is to promote and preserve the American folk art of building traditional scarecrows. To encourage this, a $1,000 prize is given for the best traditional scarecrow each year. Entries come from hundreds of miles away, as well as from local families, scout troops, church groups, and individuals.

Most entries show not only creative talent, but also a touch of humor. "Ma and Pa Kettle" were made entirely of various kitchen tools! The "Junk-food Junkie" wore a shirt and hat made entirely from wrappers of candy bars, potato chips and other snack foods. An entry by a Girl Scout troop featured crows in ballet costumes.

Some scarecrows parody movie and TV stars, such as "Vanna Whitecrow," standing in an elaborate gown next to a wheel of fortune. Other "stars" have included "Doctor Crowscare," "Bruce Scaresteen," "Marilyn Moncrow" and "Ramcrow."

Not all entries are humorous. Some are simply beautiful. One of these, "Harvest Queen," was created with dried seeds, seed pods, wheat and grasses. Though a chicken wire frame was used, the materials were so skillfully attached, the frame seemed invisible.

A complex of shops called the Peddler's Village, in Lahaska, Pennsylvania, has held a scarecrow festival and contest every autumn for more than a decade. Categories include Whirlygig, Extraordinary, Contemporary, and Traditional. Visitors cast "straw votes" for their favorite scarecrows, and the winners pick up cash prizes. The festival also includes scarecrow-building workshops, where people can make scarecrows to take home with them.

A smaller contest held in Oakley, California, is more like the old-

*Egg beaters, cheese graters, funnels, wire whisks,
and, of course, a kettle or two, are some of the kitchen tools
used to build "Ma and Pa Kettle."*

57

*Plastic garbage bags were used to create this pair of crows.*

fashioned farm-community contests. Held at Dwelly Farm's produce stand, it is open to local schools and families. There are two categories, traditional and non-traditional. The entries are on display the last two weeks of October, with small cash prizes awarded in each category.

Perhaps there are scarecrow contests in your area. Check with your chamber of commerce or local fair board. If there aren't any contests nearby, you might want to start one of your own.

Decide which categories you want. You could have Scary, Funny, Most Unusual, Best Use of Natural Materials, Traditional, or any combination of these. There could be age divisions, as well as separate categories for scarecrows made by groups and individuals. These are all ways to encourage more entries.

Where to hold the contest, how to advertise it, and what prizes you will offer are all important considerations. Tying your contest in with another event is helpful. Does your community have a harvest fair, Oktoberfest, or pumpkin patch? How about a farmer's market? These would all lend themselves well to a scarecrow contest.

There are lots of ways to put together a contest. However you decide to stage it, it's sure to be lots of fun!

*King of the cornfields*

## AUTHOR'S ACKNOWLEDGEMENTS

It is with joy and gratitude that I acknowledge all those who, in one way or another, have helped make this book a reality. Warm and loving thanks go to:

My husband, Richard, who learned to do his own laundry, took the kids on outings, and insisted we buy a computer so that "mom could write." And for so many other ways he supported and encouraged me to finish this book.

Our children, Brandi, Trevor and Cassi, who learned to do more around the house (so I could do less) and who kept their eyes open for scarecrows of every kind. Thanks also, for their patience on the rough days and their hugs. And for all the times they proudly said, "My mom is a writer."

Libby Nelson, Children's Librarian at the Fairfield-Suisun Community Library, for her enthusiasm and belief in the need for this book. She helped with research, ideas, publicity and encouragement.

The Golden Machetes, my critique group, who helped patch together the idea for this manuscript. Thanks too, for their editing, prodding and other supportive efforts.

Dorothy Kupcha Leland, my publisher, who helped me clarify and polish my work. And especially for allowing me to be involved in every step of the book's birth.

The staff and management of the Nut Tree, for the generous use of their photos, and for their support. Special thanks go to Mike Green.

Diane Wilde, for her lovely cover and wonderful drawings.
Teresa Willis and Robert Allen, for photos, time and effort on behalf of this project.

Howard Babcock, William Beers, Peter Konzak — for answering so many questions about farming, scarecrows, noisemakers, etc. And especially for sharing their knowledge with a "city girl."

Virginia and Joe Hastings, my favorite English teachers, who gave me the full "Hastings Treatment." Virginia was my ninth grade English teacher, and Joe, my journalism instructor at Napa High School from tenth grade through my senior year. They gave me the tools I use today, and laid the foundation for my career with their belief in my ability.

ABOUT THE PHOTOGRAPHERS

Bob Allen, photographer and historian, has documented many aspects of Vacaville and Solano County history. In addition to assignments for the Nut Tree, he does archival photography for the Vacaville Heritage Council and the Vacaville Museum. His work has appeared in many publications. In this book, his photos appear on pages 8, 11, 13, 16, 17, 18, 19, 27, 28, 29, 31, 32, 33, 34, 35, 39, 41, 46, 47, 49, 50, 51, 52, 53, 57, 59.

Teresa Willis is an award-winning news photographer for *The Reporter*, in Vacaville. In this book, her photos appear on pages 15, 26, 37, 42, 43, 44, 45, 55, 58 and the back cover.

BOOKS AVAILABLE FROM TOMATO ENTERPRISES

*All About Scarecrows*
By Bobi Martin. 1990. $7.95.

*The Big Tomato: A Guide to the Sacramento Region*
By Dorothy Kupcha Leland. Fourth edition, 1990. $8.95.

*Patty Reed's Doll: The Story of the Donner Party*
By Rachel K. Laurgaard. Reprint edition of the now-classic children's story about Patty Reed, an eight-year-old survivor of the Donner Party tragedy. $7.95.

*A Short History of Sacramento*
By Dorothy Kupcha Leland. 1989. $9.95.

These books are available from your local bookseller, or by direct mail. To order by mail, send price of book, appropriate tax, plus $1.50 shipping per order to:
    Tomato Enterprises
    P.O. Box 2805
    Fairfield, CA 94533
    (707) 426-3970

Bulk order discounts are available.